ROBOTS HELPING OUT

WORLD BOOK

www.worldbook.com

Co-published by agreement between Shi Tu Hui and World Book, Inc.

Shi Tu Hui
Room 1807, Block 1,
#3 West Dawang Road
Chaoyang District, Beijing 100025
P.R. China

World Book, Inc.
180 North LaSalle Street
Suite 900
Chicago, Illinois 60601
USA

Library of Congress Control Number: 2025938168

Robots
ISBN: 978-0-7166-5814-6 (set, hard cover)

Robots Helping Out
ISBN: 978-0-7166-5822-1 (hard cover)

Also available as:
ISBN: 978-0-7166-5832-0 (soft cover)
ISBN: 978-0-7166-5842-9 (e-book)

WORLD BOOK STAFF

Writer: Jeff De La Rosa

Editorial

Vice President
Tom Evans

Senior Manager, New Content
Jeff De La Rosa

Associate Manager, New Content
William D. Adams

Content Creator
Elizabeth Huyck

Proofreader
Nathalie Strassheim

Graphics and Design

Senior Visual Communications Designer
Melanie Bender

Photo Editor
Rosalia Bledsoe

ACKNOWLEDGMENTS

Cover: © 35lab/Shutterstock; © KUKA AG; © Diligent Robotics Inc.; © University of Glasgow
4-5 © PaO Studio/Shutterstock; © Diego Cervo, Shutterstock
6-7 Library of Congress; © Renault/ullstein bild/ Getty Images
8-9 © Universal Robots
10-11 © KUKA AG
12-13 © Denver Post/Getty Images; © Savioke
14-15 © Diligent Robotics Inc.
16-17 © Starship Technologies; © Nuro
18-19 © B Planet/Shutterstock
20-21 Andreas Dantz (licensed under CC BY 2.0)
22-23 © iRobot Corporation
24-25 © Robomow Friendly House; © Peter Radosa, Shutterstock
26-27 © Amazon.com; © Peng Peng, Ni Yanqiang/ Zhejiang Daily Press Group/VCG/Getty Images
28-29 © University of Glasgow
30-31 Children's National Health System
32-33 © Robin Nelson, ZUMA Press/Alamy Images; © Veebot Systems
34-35 © UVD Robots; © Aethon
36-37 © Boris Roessler, Picture Alliance/Getty Images; Jahja Jaychay (licensed under CC BY-SA 4.0)
38-39 © RIKEN
40-41 © Jens Kilian, Fraunhofer IPA; © CT Asia Robotics
42-43 © Carsten Behler, Fotogloria/UIG/Getty Images
44-45 © The University of Auckland
46-47 © sketikd, iStockPhoto

Contents

Terms defined in the glossary are in type **that looks like this** on their first appearance on any spread (two facing pages).

introduction

Do you enjoy doing chores? Whether it's taking out the garbage, lifting heavy boxes, or tightening bolts over and over again, we could all use a little help now and then to get through hard or boring tasks. People who are sick, disabled, or elderly often need extra help. Many of us are just too busy to get it all done. Could robots lend a hand? You bet!

Dinner is served! Kitchens are already full of gadgets that make cooking easier. One day, these may include fully functional robot chefs.

Why sweep the floors when a robot can do it? Robotic vacuum cleaners have caught on as household helpers.

Robots were invented to make life easier for humans, and every year more roll and ping into our lives. Many homes have a robotic vacuum cleaner or lawn mower. Factory and warehouse workers get a helping hand from robots on the job. And inventors are developing new robot helpers for almost every need you can imagine.

Someday, your life may be full of robot helpers. Their fancy robot cousins may be out building cars, fighting wars, and exploring space. But humble helpers will change our lives in small ways too.

Teamwork Makes the Dream Work

Robots and people have been sharing the work for a long time. But they have not always gotten along well.

The first **industrial robots** started working in factories in the 1960's. They were robotic arms used to put together automobiles and

Early industrial robots got some of their first jobs in automobile factories. Robots were perfect for the high workloads, heavy parts, and dangerous tasks.

other products. These brutes were heavy and powerful. They worked very quickly. Human workers had to stay out of the way to avoid being struck or crushed. Engineers often put cages around these robots to keep people away.

Human and robot workers mostly stayed in separate corners until the 1990's. Then, inventors became interested in robots that could work side by side with people. They began developing a new kind of robot, called the **collaborative robot** or **cobot.** *Collaboration* means *working together.* Cobots are built to work hand-in-hand with human workers.

DANGER:
Robot at work
Traditional industrial robots can injure people, so these robots are often placed within safety cages. The next generation of robots are being designed to work more safely alongside human beings.

ROBOT CHALLENGE

Please Be Gentle

An ordinary **industrial robot** is made of hard, heavy metal with lots of corners. Its "muscles" are electric motors. These robots are often designed to do hard work at high speed, with lots of **torque.** But a strong, fast robot can easily clobber or crush its human coworkers if they get too close.

Cobots are changing the game, with features that help them to play nice with people. First, they can modify the amount of force they use. A cobot's joints have special **sensors** that detect speed and force. If a cobot moves too fast or bumps into something, it can tell by the sensors. The cobot can ease up or stop before hurting anybody.

Cobots are also much lighter than ordinary industrial robots. A cobot arm made of lightweight aluminum and plastic will be less likely to hurt people if it bumps them. Cobots are often designed with rounded shapes, rather than sharp edges. Rounded shapes are gentler on soft human bodies. They can bump or be bumped without cutting or bruising.

Cobots are designed to work closely with people. Some cobots act differently when they sense people nearby. They might avoid going near coworkers, slow down, or even stop.

HELLO, MY NAME IS:

IIWA

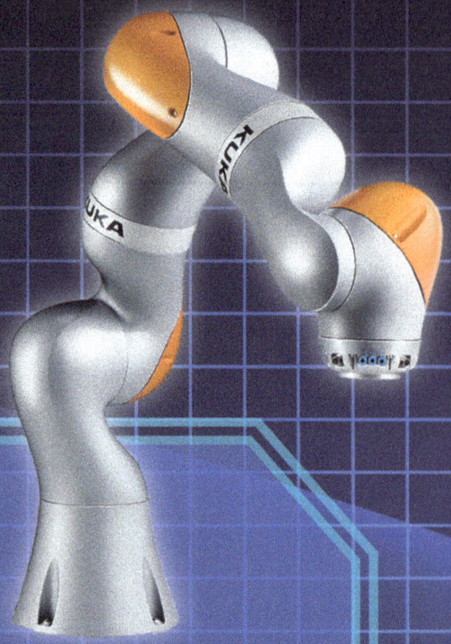

IIWA is a lightweight robotic arm with seven joints. It can bend, turn, and twist much like a human arm, only IIWA is more precise and steady. IIWA is a **cobot,** designed to work shoulder to shoulder with human beings. Touch **sensors** help it to avoid bumping coworkers. IIWA can be fitted with different **effectors** for different tasks for different tasks. It can glue car parts, hold parts for human workers, pack boxes, and even pour sodas. Human workers can teach IIWA new tasks using a tablet or by moving its arm to show it how to do a job.

AUTONOMY

MEDIUM

IIWA is good at repeating tasks and learning new ones. But like all cobots, IIWA is designed to work with humans, not on its own.

PERSONALITY

What do you expect? It's just an arm! But IIWA is quiet, works hard, and gets along with people.

WHAT'S IN A NAME?

IIWA stands for *I*ntelligent *I*ndustrial *W*ork *A*ssistant.

WEIGHT

As low as 49 pounds (22 kilograms). That's pretty light for a hardworking robot!

MAKER

KUKA robotics company of Augsburg, Germany

Special Delivery from: A. Robot

Don't feel like lugging a delivery yourself? Why not ask a robot? The first robots with office jobs were mail-delivery robots in the 1970's and 1980's. These robots carried letters and packages around large

The mail must go through
Mail-delivery robots lent a futuristic touch to some office buildings during the 1970's and 1980's.

Should you give it a tip? Some hotels use small robots to deliver towels, snacks, or other items to rooms.

office buildings. A mail-delivery robot was pretty much a rolling mailbox. It followed a set path every day, making stops around the office. At each stop, workers took their mail from slots on the robot.

In the 1990's, the spread of email meant less paper mail. Many of the old mail-delivery robots were put out of a job. But robots are still making deliveries. Hospital robots deliver food, medication, and test samples. Some hotels use robot bellhops to deliver room service, clean towels, and other items to guest rooms. And in office buildings, roaming robots bring coffee and snacks instead of memos.

HELLO, MY NAME IS:

Moxi

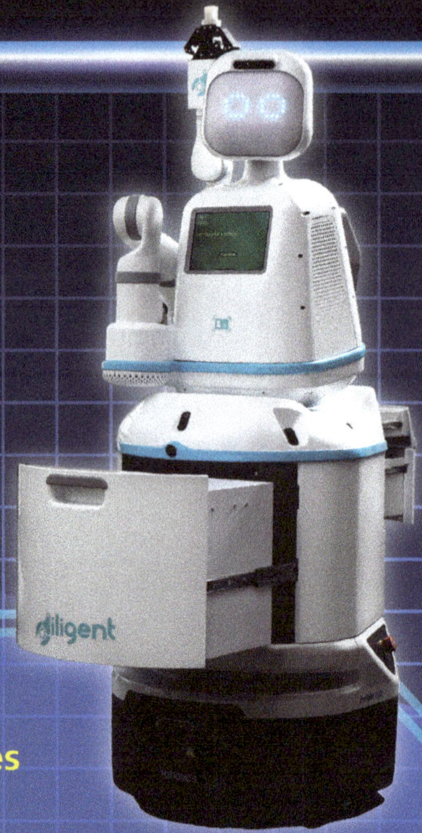

Moxi is an intelligent mobile robot designed to carry supplies and do small tasks in hospitals, freeing medical staff for more important work. Moxi rolls on wheels and can find its way around buildings. Its locking drawer can hold masks, gloves, medicines, and other supplies. Moxi's single arm allows it to open doors and operate elevators. It can also select and sort medications from store rooms. **Sensors** on the robot help it navigate and avoid bumping into people. Moxi's head sports an LED screen that can play videos, display maps, or give the robot a face to help it interact with people.

AUTONOMY

HIGH

Moxi is fully autonomous. Just tell it what to do.

MAKER

United States company Diligent Robotics

PERSONALITY

Moxi is designed to be friendly and helpful to staff and patients alike.

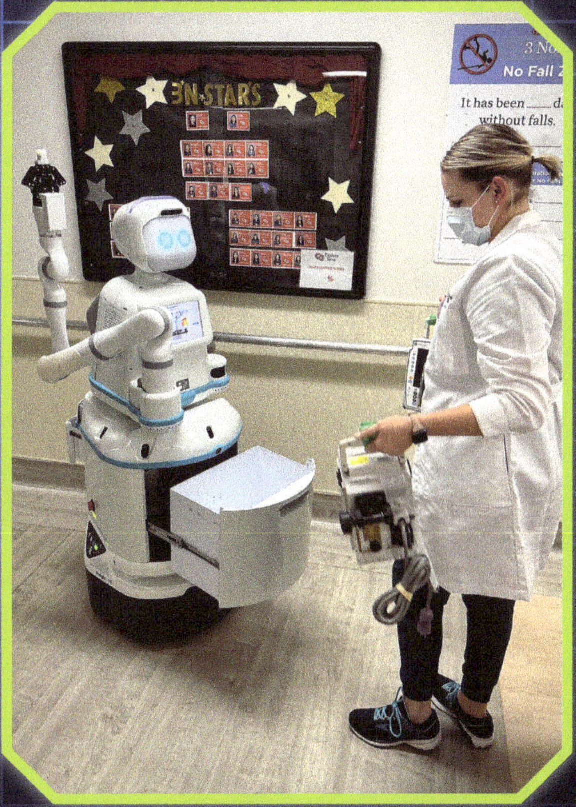

SAFE SERVICE

During the COVID-19 pandemic, Moxi could deliver supplies while keeping staff safe.

HANDY HAND

Having a hand allows Moxi to operate in environments built for people.

Takin' it to the Streets

Not all deliveries take place within a single building. Sometimes, a pizza needs to find its way from a pizzeria to a late-night party. Busy parents may need bread, milk, and eggs delivered for their hungry children. Traditionally, these deliveries were handled by human workers in cars or on bikes. But some robots are getting in on the delivery game.

The food-delivery robot Starship looks a little bit like a picnic cooler on wheels. It is built to travel city sidewalks at walking speeds. A locked, insulated holding area keeps food warm (or cold) and safe. The robot uses the

Starship-ment of deliciousness
The robot Starship can deliver food to your front door, but can't climb stairs or ring the bell.

R1 is number 1
Many self-driving car companies are still working out the kinks with their robots. But R1 is already delivering groceries to paying customers.

Global Positioning System (GPS) to find the right destination. Starship uses **sensors** and cameras to avoid pedestrians and navigate crosswalks. Behind the scenes, human operators monitor the delivery 'bots and can take over by remote control if the rolling cooler gets into a jam. Starship is racking up training miles delivering food on college campuses and in big cities.

For larger deliveries, some companies are testing self-driving delivery trucks. These are meant to drive on city streets, much like a self-driving car. Other companies are exploring delivery by package-carrying flying drones. Drones are especially useful for delivering to remote areas where roads are bad.

ROBOT CHALLENGE

Cleaning House

Compared to navigating city streets, asking a robot to clean the kitchen floor might seem like a breeze. But even this simple household chore takes some pretty clever navigation skills.

No dust bunnies here Cleaning under the bed is no problem for a low-slung robot vacuum cleaner.

A robotic floor-cleaner has to find and clean dirt. It also needs to sense and avoid obstacles—not crash into furniture, tumble down stairs, or run over the cat. The robot can't get stuck under furniture. And, it must be able to find and return to its charger when its battery runs low.

Perhaps the hardest challenge is to clean every bit of floor. Mapping out and following a good path to cover an entire room is a surprisingly complex task.

The Random Walk

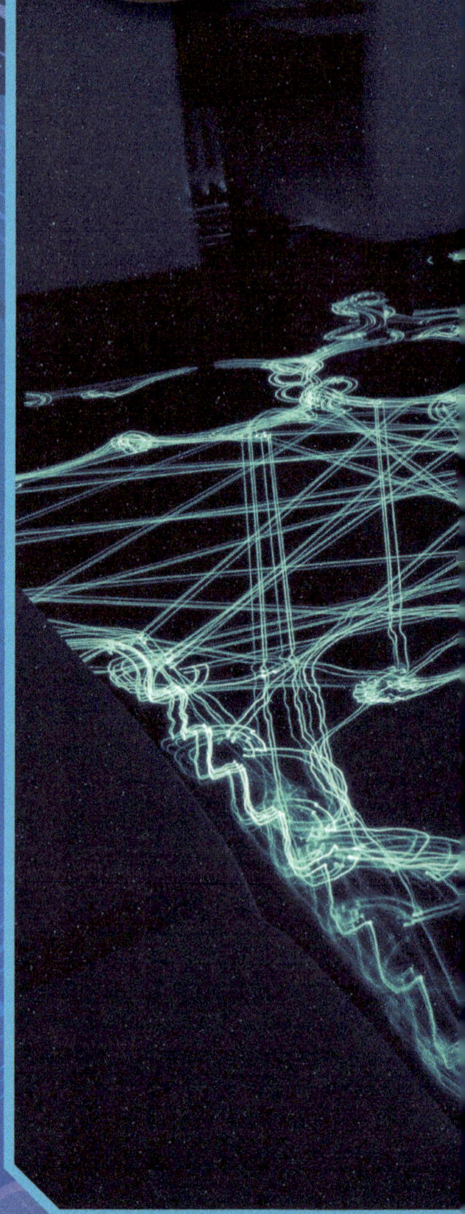

In the early 2000's, inventors made a major breakthrough in floor-cleaning robots. They realized that to cover a room, a robot does not need to have to map and follow a complex path. Instead, they built robots that make use of a strategy called the *random walk.*

A random walk involves moving according to a simple set of rules. The robot sets out in one direction. When it runs into an obstacle, it simply turns and rolls off in a new direction. The movements of such a robot may appear random. But by criss-crossing a room over and over, the robot has a good chance of covering the entire floor.

The random walk is not very efficient—the robot may clean the same spot several times just to make sure it gets everywhere once. But unlike people, robots have the time and patience to bump and bop around the living room all afternoon. They can also improve their efficiency by combining the random walk with such strategies as detecting dirt, moving in a spiral, and following walls.

Random acts of clean-ness
This time-lapse photo shows a robotic vacuum's random path around the floor. It covers many parts several times, but it manages to get every part at least once.

HELLO, MY NAME IS:

Roomba

Roomba is a robotic floor vacuum. It ranks as perhaps the most successful household robot ever. This disk-shaped dust-buster first swept into homes in 2002. Roomba combines a powerful vacuum cleaner with a rechargeable battery. **Sensors** help it navigate the obstacle course of the modern living room. **Infrared light** sensors trigger the Roomba to slow down as it approaches a wall, a piece of furniture, or other obstacle. Touch sensors stop the Roomba when it bumps into something. Newer Roomba models combine random walk techniques with room mapping and dirt sensors for more efficient cleaning. Some can even mop.

AUTONOMY

HIGH

Roomba sweeps up while the family is out working or playing.

POPULARITY

Millions of Roombas have been sold worldwide.

TURNS ON A DIME

Roomba's two wheels can spin in opposite directions. This enables the robot to turn completely around—and keeps it from getting stuck behind the couch.

CLIFFHANGER

Infrared "cliff sensors" in Roomba's bottom tell the robot when it is nearing a slope or ledge, preventing a nasty tumble down the stairs.

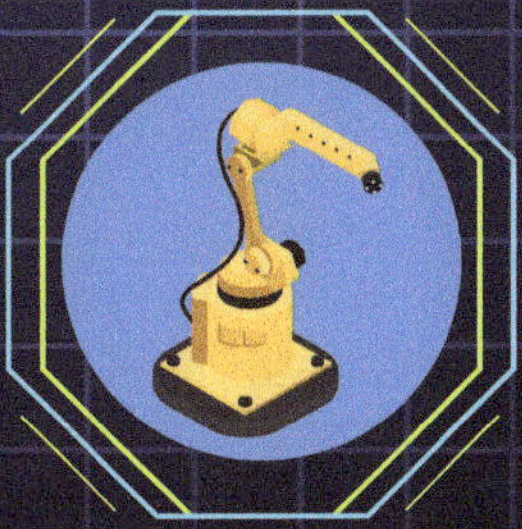

MAKER

The iRobot Corporation of Bedford, Massachusetts, in the United States

Meet the Help

Roomba and other robotic vacuums are so simple and useful that they have inspired an entire range of robot floor cleaners and groundskeepers. There are Roombalike robots that will sweep and mop hard floors, clean up pet hair, and wipe up spills.

Floors are not the only surfaces that need tidying. A rechargeable robotic lawn mower, such as Robomow, cuts the grass in much the same way robotic vacuums tend the carpet. In automatic mode, the robot wanders the yard, its twin blades trimming grass to an even

Spend less time doing yard work... A robotic lawn mower will keep your lawn in shape, tirelessly trimming on even the hottest summer day.

...and more time relaxing by the pool
Why waste time cleaning the pool? A robot can do that, too! A robotic pool cleaner scours the bottom of the pool, scrubbing away dirt and cleaning the water.

height. Some robotic lawn mowers have a **sensor** that detects moisture, so the robot stays put on wet days.

Other robot rovers are not afraid of a little moisture. Several companies sell robotic pool cleaners. These robots roam the bottoms and even the sides of swimming pools, scrubbing off gunk and filtering dirt from pool water.

Robotic Assistant, at Your Service

Some household robots aim to do more than mow or mop. They do not tackle just one tough job, as do Roomba and Robomow. Instead, these robotic assistants are eager to help out in lots of little ways.

"Alexa, where did I leave my keys?"
This puck-sized smart speaker gives access to a "virtual assistant" that can answer questions and help in other ways.

Many households already make use of a virtual assistant, such as Apple's Siri or Amazon's Alexa. These are voice-activated computer programs. They are often used through a smartphone or computerized speaker. A virtual assistant can answer questions, play music on request, and make purchases online. It can even control such household devices as lighting and locks over a computer network.

Robot butler?
The Walker robot, developed by the Chinese company UBTECH, might be able to help with many factory and household tasks.

It is also possible to host a virtual assistant or **artificial intelligence** (AI) in the body of a robot. The result is a robot helper that can fetch you a drink, order a pizza for delivery, put away toys, and remind you of your dentist appointment.

HELLO, MY NAME IS:

RoboGuide

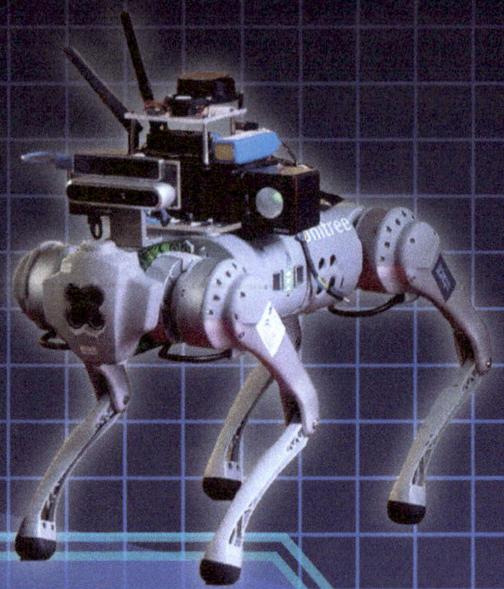

People who are blind or visually impaired sometimes use trained service dogs to help them get around. Soon, some of those dogs may be robots. Robot dogs are widely available and can be programmed for a number of different tasks. Engineers at the University of Glasgow have programmed RoboGuide to guide visually impaired visitors around a local museum. The dog 'bot leads like a service animal and can answer questions and describe objects. Robot guide dogs could also be helpful for people who are allergic to dogs or in places where dogs are not allowed.

AUTONOMY

HIGH

A human directs RoboGuide where to go, then the dog leads the way.

FOLLOW ME

RoboGuide responds to spoken commands and tugs, just like a real guide dog.

STAIR MASTER

RoboGuide can climb stairs and curbs, making it more useful than rolling robot guides.

SIZE

RoboGuide is the size of a small dog, but the programming can be used with any size of robotic dog.

FAMILIAR FRIEND

For visually impaired people used to working with dogs, a robot dog will seem familiar.

LOOKING OUT

RoboGuide is trained to recognize objects and avoid obstacles. Future models could be trained to read signs.

Dr. Robot, M.D.

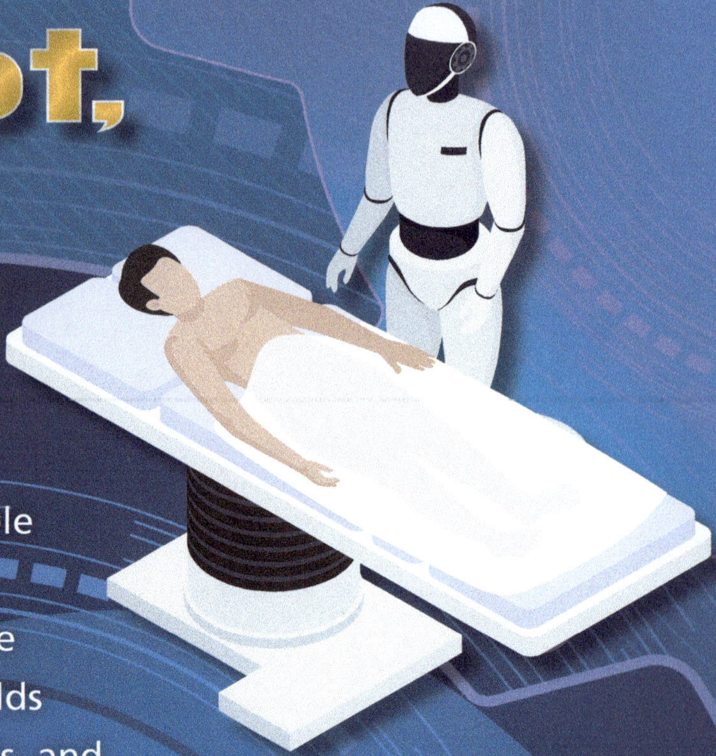

Home is not the only place where people need robot help. Imagine a busy hospital. There are injured people in need of surgery. There are sick people to feed and care for. There are medicines to be delivered. All that adds up to a lot of work for doctors, nurses, and other staff. These hardworking healers could use some robot help.

Hospitals began doing robot-assisted surgery in the 1980's. This kind of surgery makes use of robotic arms with scalpels (surgical knives) and other tools at their tips. These robot arms are remote-controlled by a surgeon, but the guiding surgeon could be in another city.

Newer robot surgeons have more **autonomy,** often aided by **artificial intelligence.** An experimental robot called STAR (*S*mart *T*issue *A*utonomous *R*obot) has been designed to help stitch up patients. In one test, STAR did better at stitching a pig's intestine than did human surgeons.

STAR surgeon
The robot STAR, shown being adjusted by a technician, has beaten human surgeons at making clean cuts and removing mock tumors, in addition to stitching.

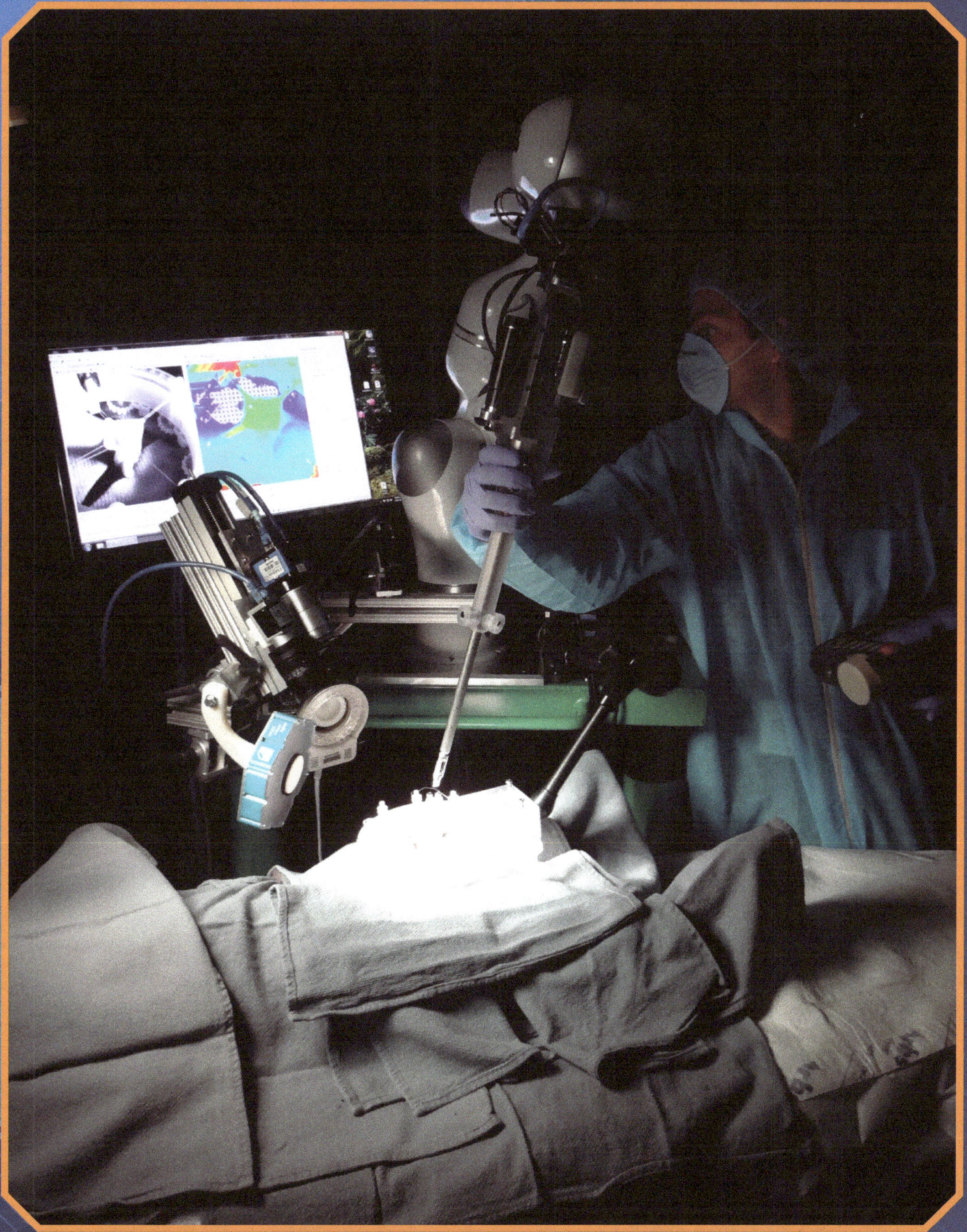

Measuring and giving out medications is another chore that robots can help with. Counting pills is boring and repetitive. Yet it must also be done with care. A single mistake can endanger a patient's life. Luckily, robots are great at doing the same thing over and over without mistakes. And, robots never get bored.

Prudent pill-packer
Robotic pharmacies reduce waste by dispensing medications before their expiration date.

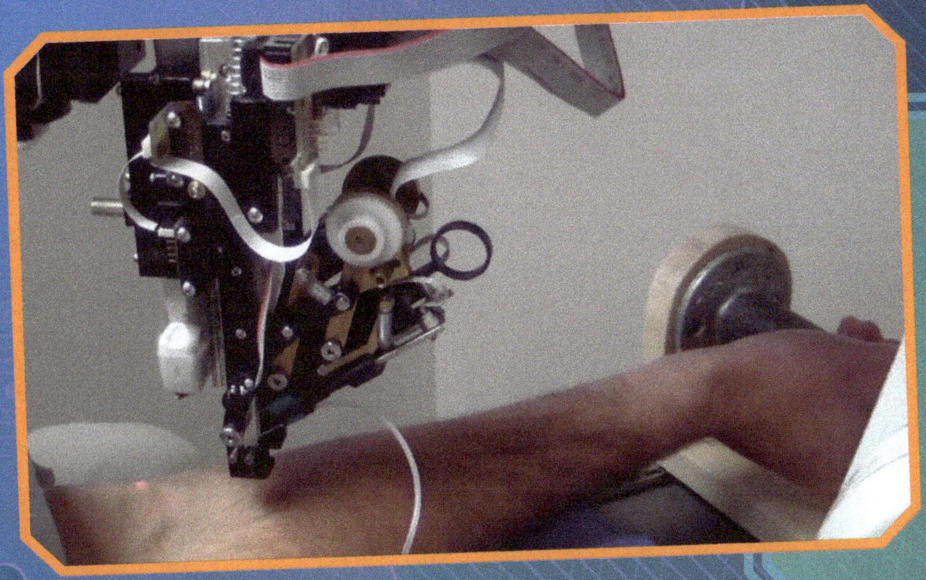

This might hurt... a little less
AI-guided robots may be better at finding a good vein, making blood draws less painful.

Many hospitals and pharmacies make use of robots to measure and give out medication. The Omnicell XR2, for example, is a robotic pharmacy designed for use in a hospital. The XR2 has a robotic arm that zips to and fro in a specially designed room, sorting, storing, and dispensing medication.

Would you let a robot stick you with a needle? The Dutch company Vitestro has designed an autonomous robot that draws blood for testing. The device uses **infrared light** and **ultrasound** to scan the patient's arm and find a good vein. It draws blood on the first try about as often as a human professional.

HELLO, MY NAME IS:

UVD

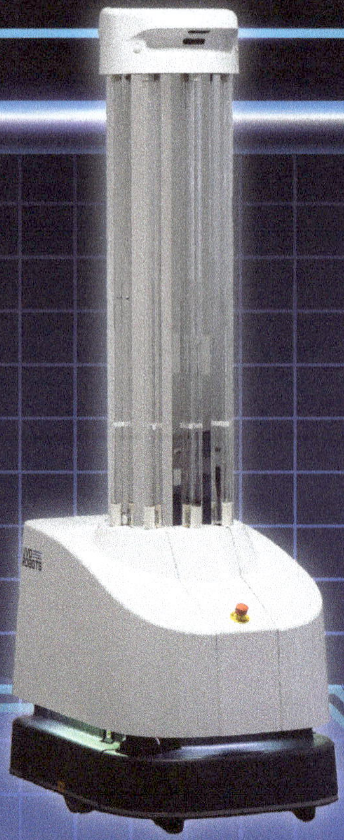

Schools, airports, and hospitals can get germy. Keeping them clean is a constant chore. One efficient way to kill germs on surfaces is with a burst of high-energy **ultraviolet** (UV-C) light. This breaks up bacteria and viruses. Unfortunately, it is also harmful to human cells. Time to send in the robots! UVD (ultraviolet disinfection) robots are rolling 'bots designed to disinfect rooms with a bank of powerful UV-C lights. Once given a route, they can disinfect hospitals, airports, schools, or other public spaces at night, when no people are around. The 'bots are equipped with proximity sensors and turn off if they sense humans in the space.

AUTONOMY

HIGH

UVD robots can be directed from an app, or can be set to run autonomously.

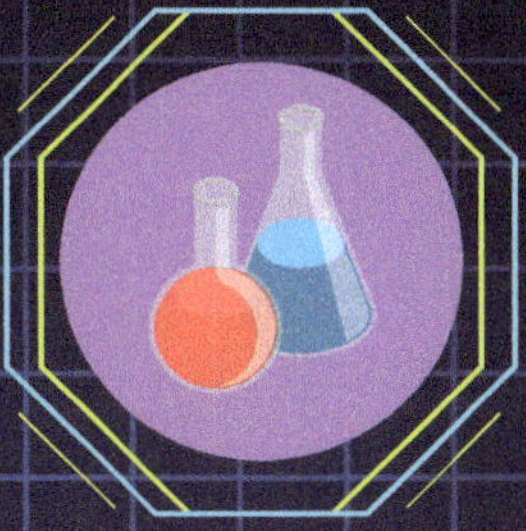

HOLD THE CHEMICALS

Disinfecting with UV light reduces the need for harsh chemical cleaners, which can cause problems of their own.

ONE JOB

UVD has just one job; disinfecting spaces.

MAKER

Blue Ocean Robotics of Denmark has made some of the first UVD robots.

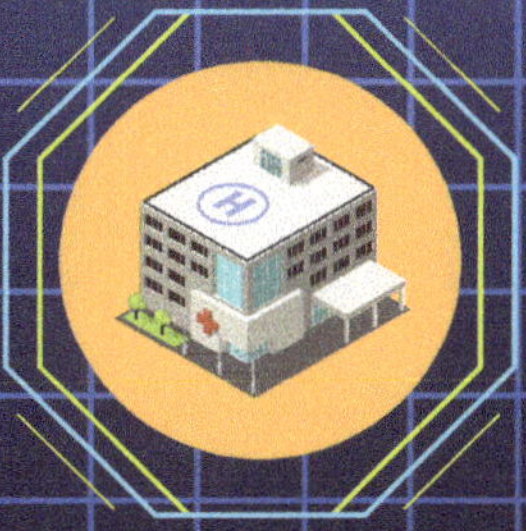

HOSPITALS PENDING

UVD robots are still being tested to see if they are powerful enough for use in hospitals, possibly together with other disinfection methods.

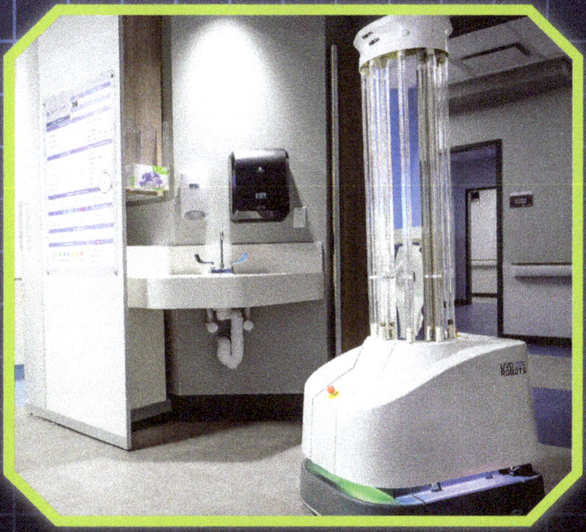

Don't Forget Grandma and Grandpa

You may not think the elderly would be first to join the robot revolution. But grandma and grandpa may get help from a robot before you do.

As people age, they begin to lose strength. It gets harder to move. Old people may also need help remembering things. Robots are ready to lend a hand.

Pepper is a robot being used in some nursing homes to interact with elderly residents and lead games and activities.

Dinsow Mini, by the Thai company CT Asia Robotics, is a companion for the elderly. It can remind them to take medication, show workout videos, and play mind-strengthening games.

The need for robot help for the elderly is especially strong in Japan. The country has a shortage of workers to take care of its aging population. Some hope that robots could help with lifting, exercising, and even playing with the elderly. So far, care robots still have a ways to go—staff often spend more time caring for the robot. But as robots and **artificial intelligence** improve, that may change.

The most helpful robots so far have been **social robots,** which can play music, tell jokes, connect residents on video chat, and remind them to take pills or do other tasks. Some can also monitor older people's sleep, alerting caregivers when they are awake.

HELLO, MY NAME IS:

Robear

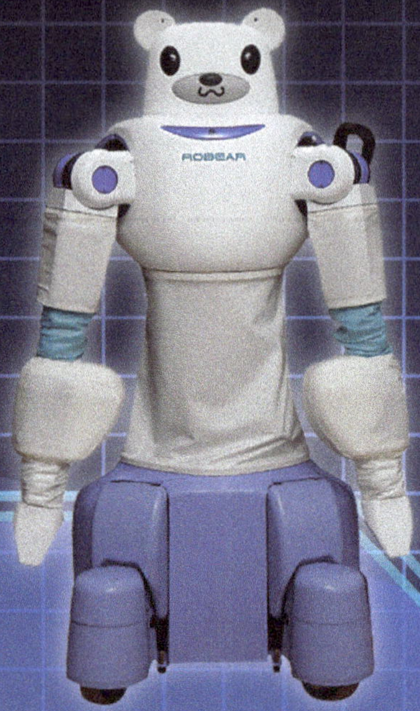

Lifting people who can no longer lift themselves can be a major struggle for nurses and other caregivers. Robear was one of several robots that have been designed to lift patients into or out of bed or help them stand up. Robear had a cute face and **torque,** touch, and other **sensors** that helped it keep its grip gentle. Robear also had a soft covering to keep from bumping and bruising its precious cargo. Robear was built in 2015, but never actually used in a care facility. The robot itself demanded too much staff time to set up, and was not very versatile. But roboticists learned a lot from the experiment.

AUTONOMY

HIGH

Robear responded to voice commands and touch.

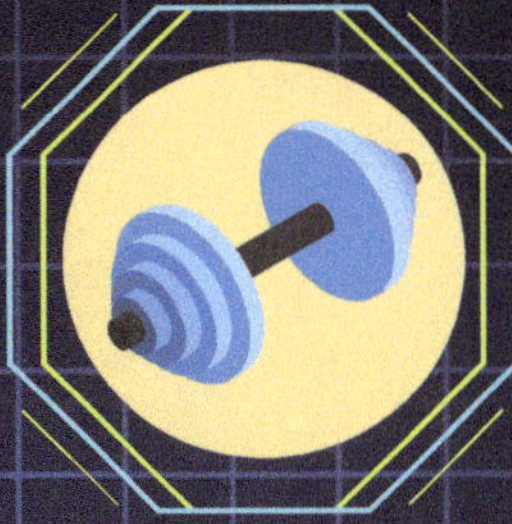

STRENGTH

Robear could lift about 180 pounds (80 kilograms).

CUTENESS

Robear's space bear face was designed to make the big machine seem friendly.

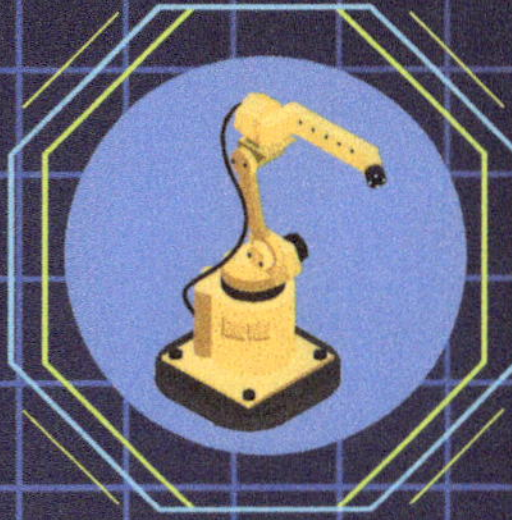

MAKER

Robear was a prototype developed at the Japanese research institute RIKEN.

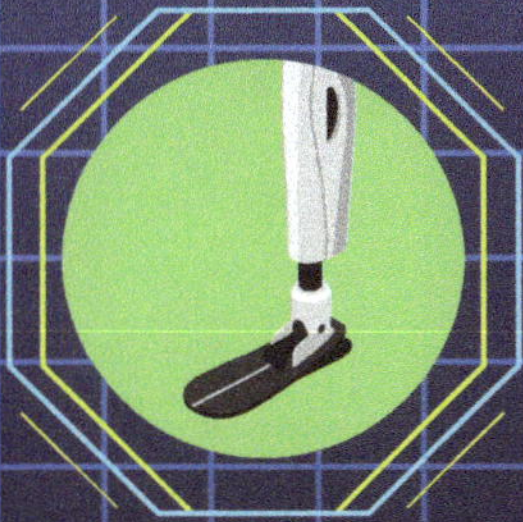

A LITTLE LEGROOM, PLEASE

Robear could steady itself on *retractable* legs (legs that can draw back into its body).

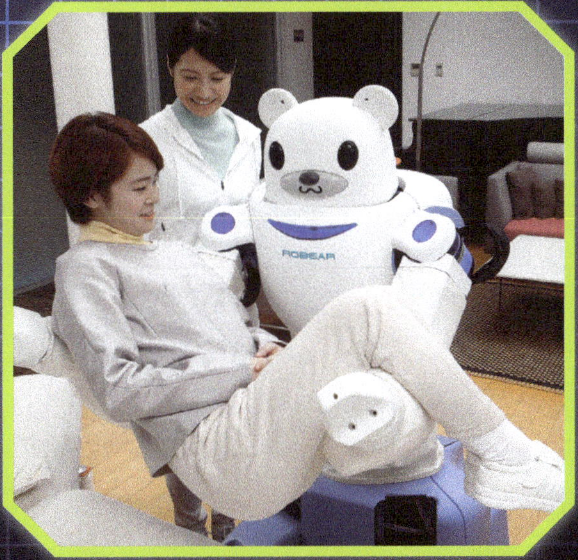

ROBOT RISK

Growing Old Among Robots

Robots can help the elderly by making daily tasks easier. Some people think they can also help with another problem: loneliness. Older people who are lonely may be more likely to suffer

A robotic assistant may be a great help to the elderly or disabled, enabling them to keep their independence.

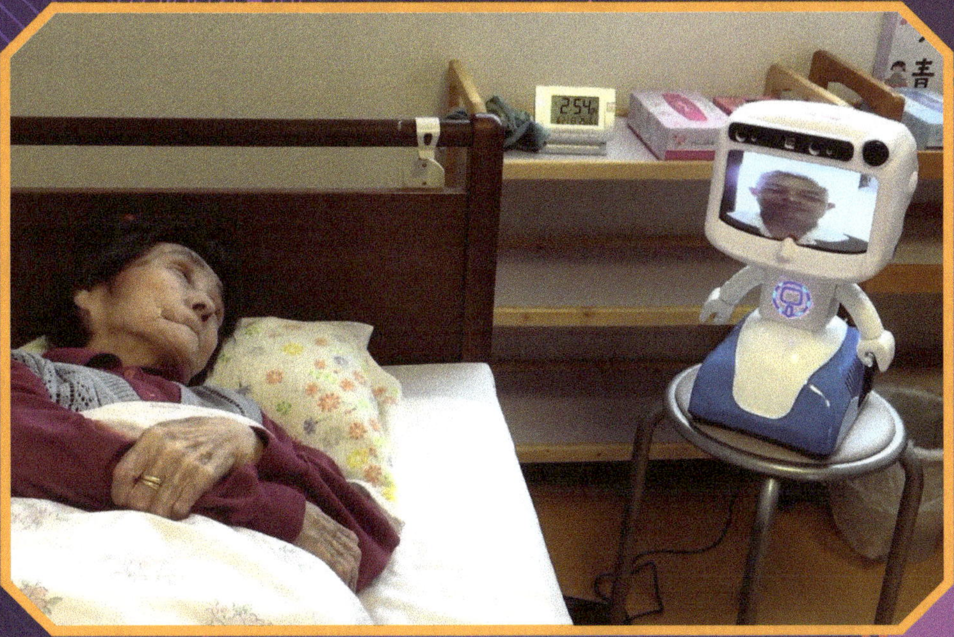

Almost like being there (but not quite)
Robots can help us video chat with elderly loved ones. But with robots taking care of them, will we visit them less?

a number of health problems, including depression. This makes loneliness a major problem in elder care.

Some robots aim to help relieve loneliness. Robots like Dinsow, for example, help people to connect with their family and others through video chats and social media. Nursing homes have also tried using robotic pets to ease their residents' loneliness.

But some experts worry that robots could make the elderly feel more lonely. They think it is important for the elderly to interact with other human beings, not just robots. They worry that older people cared for by robots will have even less contact with humans.

Robot Therapy

Robots do not just care for the sick and injured. They can also help them to get better. **Therapy robots** help people heal in body and mind.

HAL is a therapy robot that can help people learn to walk again. It is for people who have suffered an injury to the spine or a stroke. HAL stands for *H*ybrid *A*ssistive *L*imb. HAL does not look like an ordinary robot. It is a **powered exoskeleton**—a kind of robotic framework or suit that can be worn by a person. When a person tries to walk, the brain sends signals to his or her arms and legs. HAL can detect these signals. Motors in HAL's joints help the person to move.

HAL, a robotic exoskeleton by the Japanese company Cyberdine, helps to move and support people recovering from an injury or stroke.

Other robots are designed to help with mental health. They are often powered by **artificial intelligence** programs. These robots are designed to help patients deal with emotions and practice storytelling and other skills.

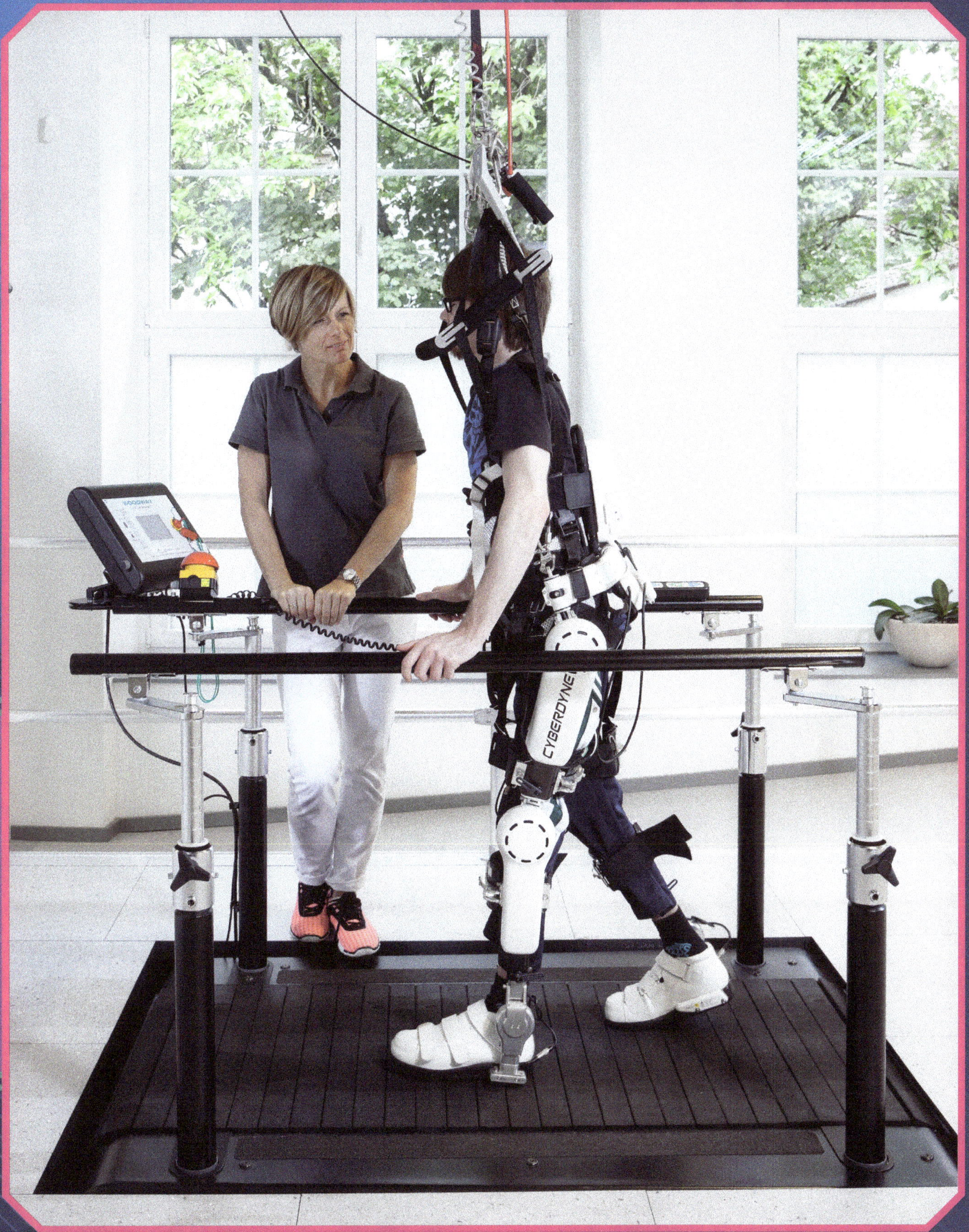
CYBERDYNE

HELLO, MY NAME IS:

Paro

You probably don't think of robots as being cuddly. But the **therapy robot** Paro is a world-class snuggler. Paro is designed to look like one of nature's most adorable creatures: the baby harp seal. This fuzzy robot loves to be petted. Stroke its plush fur, and Paro will wiggle its tail and coo with delight. Paro aims to be more than just a stuffed toy, however. Petting Paro can help calm patients, caregivers, and others in a hospital or nursing home. People often use dogs and cats for this purpose. But as a robot, Paro can go where living therapy animals aren't allowed.

AUTONOMY

HIGH

Paro learns from a person's behavior, figuring out what it takes to get more snuggles.

CUTENESS

Is anything cuter than a baby seal?

MOBILITY

Paro is more lapdog than workhorse.

MAKER

The Japanese research institute AIST developed Paro.

HEY, PARO!

Paro responds to sound, and each Paro can learn its own name.

EASY TO CARE FOR

Paro is a good companion for people who suffer from *dementia* (mental decline due to age). They may not be able to care for a living pet.

Hands-On Robotics

Want to get started making robots? Jump right in!

Tech Camps

If you'd like to spend the summer hanging out with robots, many universities, colleges, and science museums hold robotics summer camps, both in-person and online. At robotics camps, you will learn how to program and design your very own robot. There are usually divisions for different ages and interests, including games and digital arts. Check around for ones in your area.

Some education companies run summer camps focused on robotics and coding. iD Tech Camps, Engineering for Kids, and Code Ninjas are just a few. The iD Tech Camps hold summer sessions all over the United States and in some other countries, such as China. Coding is a big part of the camp experience. Working in teams, campers program their robots to do different tasks. They choose the appropriate sensors and effectors, depending on those tasks. When all the robots are ready, the teams take part in a robotics competition. Many youth organizations, such as the Scouts and 4-H, offer robotics camps as well.

Students at tech camps learn how to build and program different kinds of robots.

Look online for camps in your area, or check out the list at the NASA Robotics Alliance website.

Also check out:

- Smithsonian Institution (Washington, D.C.)
- Carnegie Mellon Robotics Academy Summer of Learning
- FIRST Robotics
- NASA Robotics Alliance

Or ask at your local school, library, or maker space.

Glossary

artificial intelligence (AI) the ability of a computer system to process information in a manner similar to human thought or to exhibit humanlike behavior.

autonomy the degree to which a robot can make decisions without input from a human operator to achieve a goal.

collaborative robot (cobot) an industrial robot designed to work closely with people and share workspaces with them.

effector the part of the robot's body, such as a wheel or a gripper, that interacts with the environment to perform an action.

Global Positioning System (GPS) a worldwide navigation system that makes use of radio signals broadcast by satellites.

industrial robot a robot that works in a factory to help create a product.

infrared light a low-energy, invisible form of light.

powered exoskeleton a kind of robotic framework or suit that can be worn by a person.

sensor a device that takes in information from the outside world and translates it into code.

social robot a robot whose primary job is to interact with people.

therapy robot a robot designed to help people heal in body or mind.

torque turning force.

ultrasound sound waves that are too high-pitched for people to hear.

ultraviolet (UV) a high-energy, short-wavelength form of light, invisible to humans. Ultraviolet can be divided into A, B, and C bands.

Index

www.ingramcontent.com/pod-product-compliance
Lightning Source LLC
LaVergne TN
LVHW070858210926
848482LV00016B/968
9780716658320